How to...
BE PRIME
MINISTER

How to... BE PRIME MINISTER

By ADAM HIBBERT

Illustrated by
Tim Benton

OXFORD
UNIVERSITY PRESS

Pour ma cherie, Clare - A.H.

OXFORD
UNIVERSITY PRESS

Great Clarendon Street, Oxford OX2 6DP

Oxford University Press is a department of the University of Oxford.
It furthers the University's objective of excellence in research, scholarship,
and education by publishing worldwide in

Oxford New York

Auckland Bangkok Buenos Aires Cape Town Chennai
Dar es Salaam Delhi Hong Kong Istanbul Karachi Kolkata
Kuala Lumpur Madrid Melbourne Mexico City Mumbai Nairobi
São Paulo Shanghai Taipei Tokyo Toronto

Series devised by Hazel Richardson

First published in 2001

British Library Cataloguing in Publication Data available

ISBN 0-19-910797-1

3 5 7 9 10 8 6 4 2

Printed in Great Britain by
Cox & Wyman Ltd, Reading, Berkshire

Contents

SO YOU WANT TO BE PRIME MINISTER?

So, you're thinking of competing for a job with some of the longest working hours, greatest stress and least popularity of all the jobs on the planet! Are you mad? Of course not. Before your family ask you to see a doctor, remind them that being Prime Minister is probably the top job in the country. Apart from the fancy houses, free flights, chauffeur-driven limo and other perks of the job, you also get to tell everyone in Britain what to do (apart from the Queen, of course). Then again, your parents might not approve when they realize you'll be in charge of *them*!

So being PM could be fun, despite all the hard work. But what are your chances of getting the job? They would be about one in ten million or so, if the PM was elected in the same way as lottery winners.

But fortunately, not everyone wants to be PM, and only a few have the right talents for the job. If you can answer yes to these three questions, you may be one of those few:

1. Am I good at getting my own way?
2. Could I handle working 18 hours a day for weeks at a time?
3. Do I really, really want to change the world?

Of course, some PMs haven't passed even these simple tests. For example, it will be up to you how much or how little you choose to work. In 2000, the PM spent a long time deciding whether to work after the birth of his fourth child.

In the end, he took a week off to be with his family. (Although he was always kept informed of what was going on and always had to be available, just in case he was needed.)

One thing is essential though – you must know your way around. British democracy has evolved from simple beginnings almost 1000 years ago, layer upon layer, and has ended up a bit complicated. This book will tell you how it all fits together, who does what, and how a future Prime Minister might go about plotting a glorious career! Along the way you will find out about:

- how local government works and what it does
-  how to get the British public to vote for you
-  PMs of the past – their successes and failures
-  what government is and what it does
-  the mysterious world of Whitehall
-  political parties – their names and what they say they believe in

CHAPTER ONE

HOW DOES THE GOVERNMENT WORK?

Let's have a quick look at the important parts of government – what it is, what it does, and the people involved. Don't worry if some of these bits seem too complicated – we'll come back to them later when we're planning your climb up the career ladder! The government is a complex organization, and sometimes looks very old-fashioned, but there is a simple idea behind it all – citizens rule!

Citizens

Because the UK is a democracy, you're the boss already! At least, you will be when you're old enough to vote. You don't have to live in a city to be a citizen; the word just means that you belong to this society, and have a role to play in its decisions and responsibilities. Citizens help run the country in two important ways.

- With your vote, you will take part in choosing who governs the country.
- With your taxes, you will help pay for all the government's work.

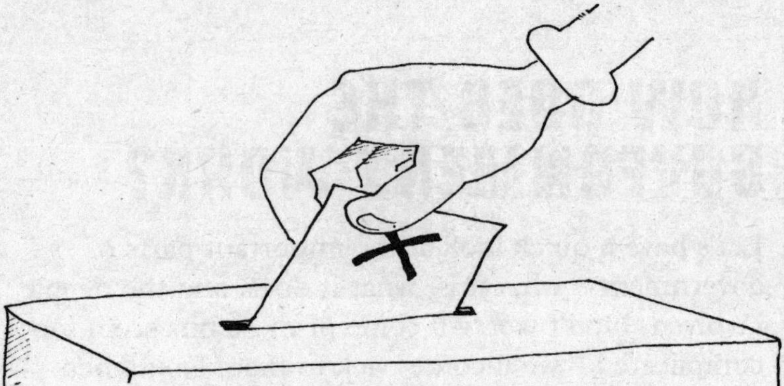

Democracy means that the grown-ups (everyone over the age of 18) in a country choose people to represent them (representatives) by voting. At elections, each citizen can go to a local voting centre, and mark a choice on a voting slip. The votes are added together, and the most popular person wins.

There are elections to different parts of government, but the most important one is the General Election. There has to be a General Election at least once in every five years. The PM may choose to call one early, to cash in on momentary popularity. An early election can also be forced on the PM if a majority of MPs gang up on the government and stop it getting its way. Either way, it's the Queen who makes the announcement. The winners in General Elections become Members of Parliament (MPs).

MPs are the men and women who take the big decisions about running the country. They debate and vote on issues in the House of Commons.

Each MP represents the voters in their area, called a constituency, or 'seat'. The UK is currently divided into 659 constituencies, each with about 90,000 voters, with one MP per constituency.

MPs try to make sure that their voters are well looked after by the government, and that problems are cleared up. They can do this by campaigning personally, or by raising issues in debates in the House of Commons.

Almost all MPs are members of a political party.

Parties aren't just for birthdays! A party is a group of people. Everyone has different ideas about how to run the country, but some people agree on enough ideas to work as a team (or'party'). This way they stand more chance of seeing their ideas put into practice. The two biggest parties have about 350,000 members each, down from over a million each in the early 1980's. The members choose a leader and suggest which members should try to become MPs.

There are three big parties in the UK. They each explain their ideas in a manifesto, which is a sort of list of promises to the voters. The details change at each election, but there are basic ideas for each party which stay the same.

MPs meet in the House of Commons, which is a large wood-panelled room in the Palace of Westminster, in London. Here MPs debate new ideas and vote to decide whether or not to put these ideas into practice. The party with the most MPs after an election can usually outvote all the other parties, so its ideas will be the ones that win in debates. This party forms a government, and decides which jobs to tackle over the next few years. The government can make a difference to citizens' lives in two ways.

- It makes new laws by winning a vote among MPs in the House of Commons.
- It runs the different departments of government according to the party's ideas.

Know your Palace!

The Palace of Westminster was the site of the earliest parliaments. It's famous around the world for the clock tower at the northern end, known as Big Ben. Big Ben is actually the main bell. It strikes the note E, and is broadcast live all over the country on New Year's Eve.

The Palace contains the House of Lords and the House of Commons, various committee rooms, meeting rooms, offices, tearooms, restaurants, and bars, as well as a hairdresser, a clinic, a florist and special spaces for journalists. There is even a small gym.

Parliament

The House of Commons is the most important part of Parliament, but not the only one! Every decision it takes has to be approved by two other parts – the Queen, and the House of Lords. This may seem a bit old-fashioned, but it sometimes helps to prevent the House of Commons doing something silly.

The Queen is the UK's official Head of State. The Queen never opposes the House of Commons, but has to give her permission for it to act.

The House of Lords is similar to the House of Commons, except that its members are not elected by voters – they include senior bishops and judges, and people the Queen has made into nobles on the advice of her prime ministers, past and present. The House of Lords can't stop the House of Commons, but it can raise problems and cause serious delays.

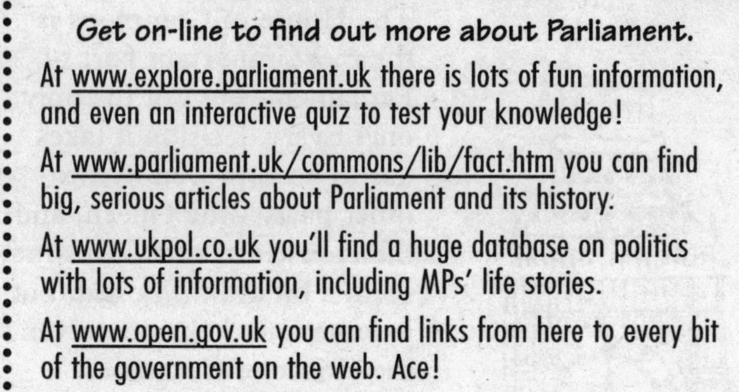

wwWestminster!

Get on-line to find out more about Parliament.

At www.explore.parliament.uk there is lots of fun information, and even an interactive quiz to test your knowledge!

At www.parliament.uk/commons/lib/fact.htm you can find big, serious articles about Parliament and its history.

At www.ukpol.co.uk you'll find a huge database on politics with lots of information, including MPs' life stories.

At www.open.gov.uk you can find links from here to every bit of the government on the web. Ace!

The Prime Minister

The PM is the leader of the party in government – the party with the most MPs in the House of Commons. The PM decides which MPs should be ministers, to run each department of government. Together with a few other advisers, this group of MPs forms a special committee, called the Cabinet. The Cabinet is the PM's top team, meeting at Downing Street to work out what the government should be doing each week.

TOP SECRET

The Cabinet is the Prime Minister's top team. There are usually around twenty Cabinet members and this small team helps the PM to run the government.

Cabinet meetings are always held in private – they are top secret. MPs in the House of Commons are presented with Cabinet decisions and asked to support them. Most MPs in the Prime Minister's own party will go along with the Cabinet's ideas, because they are all on the same side. MPs from other parties usually have objections and they often vote against the Cabinet's proposals. However, because the government is run by the party which has the most MPs, the Cabinet usually gets its way. The parties in opposition may be able to encourage the government to make some small changes if enough of them agree upon a certain point.

Cabinet members are usually senior MPs from the Prime Minister's party. They may be old friends of the PM, people who helped them win the leadership competition in the first place. Cabinet ministers may be chosen because they are experts in a particular field, say, economics.

MPs in government are in charge of a big organization, the state. The state includes all the people, buildings and equipment that the government needs to do its job.

The state employs lots of people you see in everyday life: doctors and nurses, teachers, police officers, soldiers, and so on.

It also employs lots of people you hardly ever see at work: judges and lawyers, economists, accountants, researchers, spies, diplomats, and the office staff who make sure that it all runs as smoothly as possible, often known as civil servants.

That all adds up to over a million employees, and costs £300,000,000,000 each year!

Europe

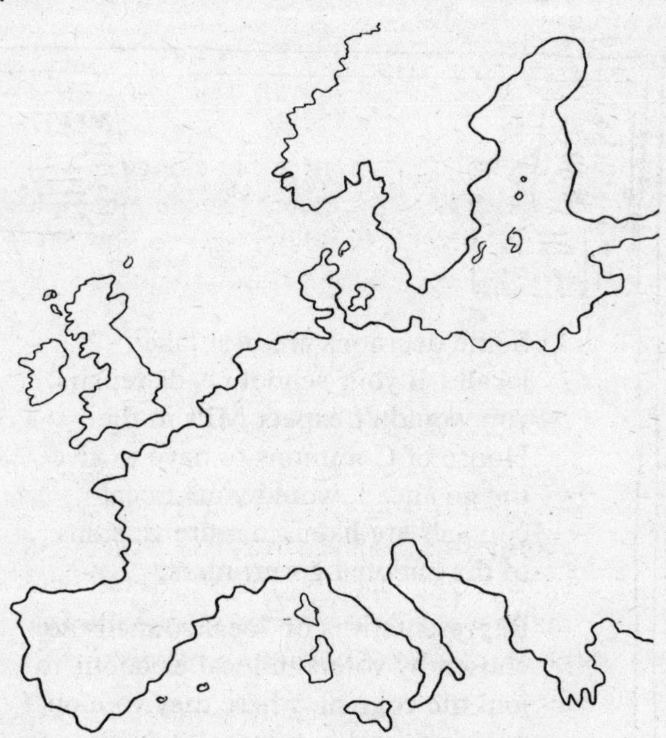

The UK is a member of the European Union. The Union was started after World War II, as a way to help businesses across Europe work together. It has now expanded to include rules about citizens' rights, amongst other things. The EU is run by commissioners, under orders from a Council of Ministers, and a European Parliament.

Each country elects Members of the European Parliament (MEPs) to represent voters in the Union. MEPs spend a lot of their time in Belgium, which isn't necessarily a bad thing (they have all the best chocolate!).

Some decisions are best taken locally. If your school needs repairs, you wouldn't expect MPs in the House of Commons to have to give the go ahead, would you? Local councils are like miniature versions of the national government.

Representatives on local councils are chosen by voters in local elections to join the council, where they vote on decisions and give instructions to local council employees. Everything from schools to community centres, housing and rubbish collection is organized locally.

Regional government

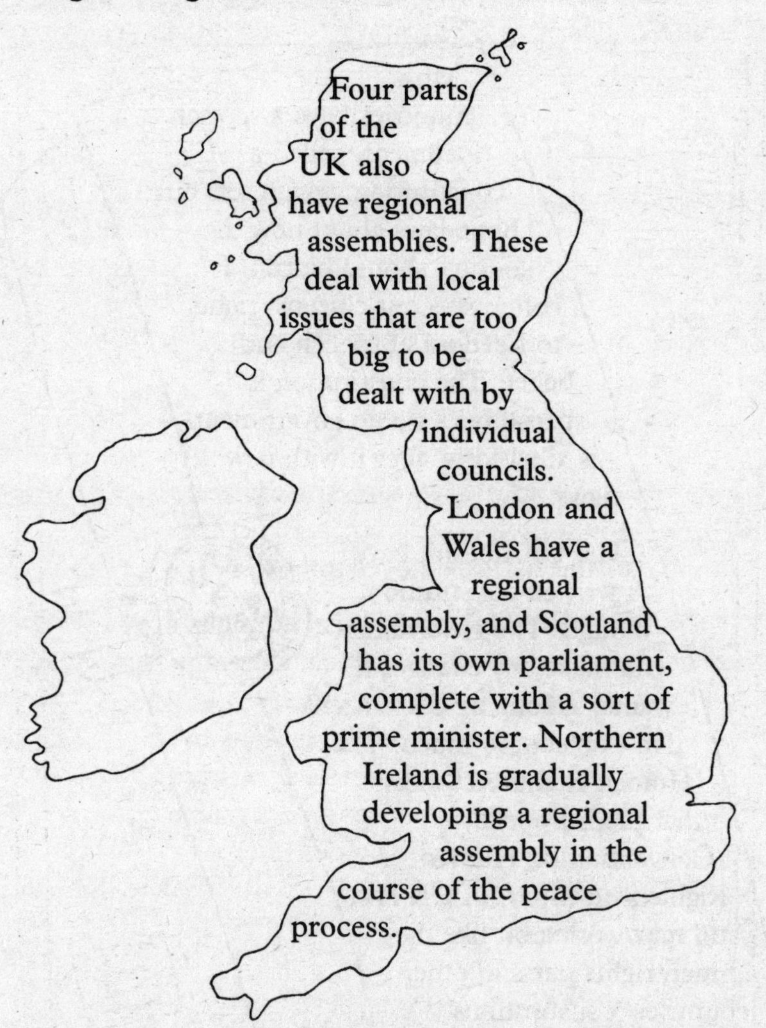

All four were created at the end of the 1990s. The process was known as 'devolution' (giving power away from the centre). The 'devolved assemblies' are still at an experimental stage – nobody knows quite how they will look when they have settled down.

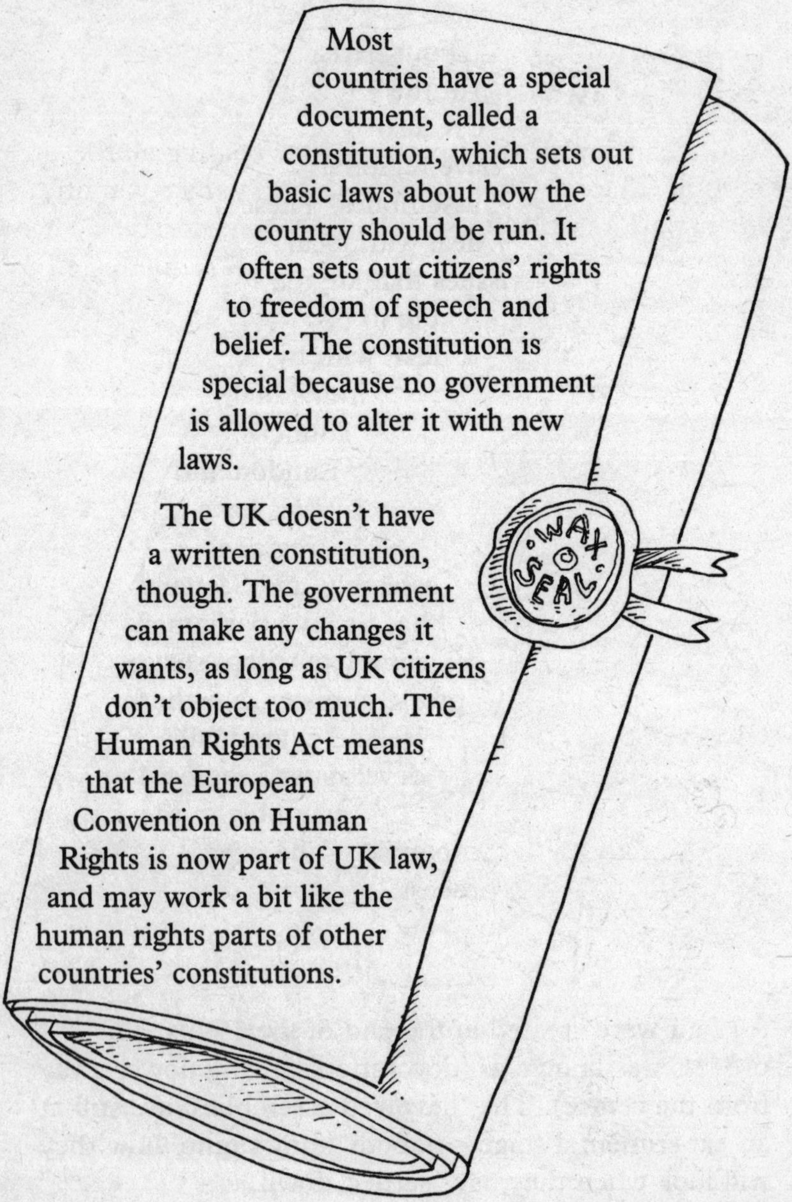

Most countries have a special document, called a constitution, which sets out basic laws about how the country should be run. It often sets out citizens' rights to freedom of speech and belief. The constitution is special because no government is allowed to alter it with new laws.

The UK doesn't have a written constitution, though. The government can make any changes it wants, as long as UK citizens don't object too much. The Human Rights Act means that the European Convention on Human Rights is now part of UK law, and may work a bit like the human rights parts of other countries' constitutions.

WAX SEAL

CHAPTER TWO

GETTING STARTED

Your head is probably spinning after reading all the stuff in Chapter One! But if that didn't scare you off, it's about time you began planning your glorious political career. Bear in mind that if you're interested in one of the major parties, youth is a big advantage. The average age of a party member in the main parties hovers around the 60-year-old mark. If you want to make a big impression, start out while you're less than a quarter of that age!

What sort of person becomes PM?

Since World War II, all sorts of people have become PM.

- The first five were all from private schools, all went to Oxford University, and four of them were from aristocratic families.
- Since Harold Wilson in the 1960s, most PMs have been from state schools, though three more of these studied at Oxford University. All of them were MPs well before their 40th birthday (Winston Churchill became an MP when he was 25). Margaret Thatcher, who was Prime Minister from 1979 to 1990, is the only woman to have held the job ... so far!

- In the 1997 elections, 126 MPs who were elected had been teachers or lecturers; 64 were lawyers, 55 had once been manual labourers.
- Out of the 659 MPs elected in 1997, eight were under the age of 30 and 120 were women.
- About half of the Liberal and Conservative MPs who were elected in 1997 were educated at a private school, but only one in 16 of Labour MPs enjoyed that privilege.

If there's a political party whose ideas you agree with, look up their address in the phone book and write to them, or look on their website. It might be best to warn your parents of your plans first – you never know, they may have a few words of useful advice! The party will put you in touch with members in your area, and you'll soon find that you can make yourself useful.

Most parties have a special group for young members, where you can make friends and learn about the party's ideas and history. Take this seriously, and make a point of being charming (though not slimy!) to everyone you meet – you never know which of them will grow up to be important. If you have to argue with another member whose ideas you think are stupid, try to be polite, generous, and friendly even after the debate, however you really feel about them!

Go it alone?

What if you don't agree with the policies of the main parties? There are alternatives. The Green Party is for environmentalists, and it's beginning to win some seats on councils and in Europe. There are smaller parties active around other special issues too, and parties which strive for independence for their areas in Scotland and Wales. (For obvious reasons, though, if you want to be the British PM, a career in Scottish or Welsh politics will have certain limitations!)

But it may be that you just have to start your own party. Think carefully about what's wrong with the existing parties. Which bits do you like? Would your new ideas appeal to the general public? Start to sketch out your party's policies. Don't announce your new party straight away – you'll need to find lots of political friends, and plenty of cash! If you can bring all these things together, and agree on a plan with like-minded politicians, you may be in business.

But before you rush off to print your manifesto, bear in mind that the only new parties to be formed in the last 200 years that have made it in Parliament have involved experienced MPs from the very beginning.

It's party time!

Let's say for the sake of argument you have decided to join one of the three main parties. The advantage of joining a big party is you'll get to meet the people you will have to impress in order to become PM. Big parties have a party conference each year, normally in the early autumn, and you can really get a head start if you pop up in front of 6000 party members there and give a good speech. You might even get on the telly! The older members are always pleased to know that their ideas will live on in your generation, and they'll love you for it.

Canvassing

Canvassing for an election is a good way to start on a political career. All parties have small groups organized in each area of the country. These groups pick someone who they think can promote their beliefs to be their candidate for an election. Talk to your local group about your party's manifesto (the list of promises it makes to the voters) and work out how to explain its main points to people.

A canvasser is someone who convinces people to vote for their party's candidate. As a canvasser, you'll end up with sore feet from walking from door to door around your neighbourhood, asking people if they will support your candidate, and trying to show them why your party's ideas are better than the other parties' ideas. Some people won't talk to you, and others might not like the party you represent. Whatever happens, you'll soon learn how best to argue your case.

It sounds boring to tramp around, quite often on rainy nights, knocking on doors where people will normally not be very pleased to see you. But there is no better way to understand politics. Canvassing, like debating, lets you test your ideas in arguments with other people, who you expect to win over to voting for your gang. Unlike debating, though, it's not a game! If you do your job well, you might actually cause a change in people's real lives. For this reason, canvassing is much more no-nonsense than debates and idle chat among friends.

As with most things, it'll be the moments when you fail that provide you with your best chances for learning important lessons. If you can't get through to a potential voter, and trip up over a fact, for example, a bit of history you don't know about, you might feel like giving up. But that's where being in a party helps.

Other people who share your ideas probably have an answer to the problem you're facing. You might have answers for things they are having trouble with. If you are canvassing with a team of people, you can all support and learn a lot from each other.

Don't forget to pay attention to those fellow canvassers. If you're going to be a leader, you'd better learn fast how to inspire people and keep them slogging away - pretty soon, you'll want them to be persuading people to vote for you!

Student politics

To get your first chance to be a candidate, you will probably have to go to college or university. University politics can be weird, but it offers people a rare chance to run big organizations before they are 20 years old. For example, each university has a branch of the National Union of Students (NUS). Getting elected to a position running the branch can be quite easy if you can persuade some student friends to vote for you. You take a year out from studying to do the job. At the same time you can impress your fellow politicians. If you're successful, you can try to get elected to the national committee, or even become president of the national union itself. NUS presidents often become MPs soon after leaving university.

Local councils

Whether or not you tried student politics, local councils are another good first step. Work hard for your local party, make friends with the other dedicated members, and after a while you can suggest that they choose you as their candidate in a local election. It will give you very useful experience of handling the press, running an election campaign, and – who knows – you may even win the election!

As a councillor, you won't get paid, but you will see how an important part of government works. Local councils are similar to national government. There are elected leaders and there are employed officials who try to make sure that your decisions are carried out. At this point you'll begin to see how a party's manifesto is turned into actions. Sometimes you'll find that you can't achieve what you want. You'll also learn to spot when the obstacles are real, and when they're just other people deliberately trying to stop you achieving success.

It's lonely at the top

Stop right now if you expect to be loved for your troubles. Politicians are often very unpopular, particularly because they have to find compromises between their hopes and those of other groups. Few other careers have been so despised throughout history.

'The most successful politician is he who says what everybody is thinking most often and in the loudest voice.'
Theodore Roosevelt
(1858–1919)

'He knows nothing; and he thinks he knows everything. That points clearly to a political career.'
George Bernard Shaw
(1856–1950)

'Since a politician never believes what he says, he is surprised when others believe him.'
Charles de Gaulle
(1890–1970)

MPs

Whichever way you do it, through student politics or local elections, or by another route entirely, you'll have to become an MP sooner or later if you ever want to take a shot at becoming Prime Minister. Most MPs are elected at General Elections, where people from all over the country turn out to vote. Some are elected at by-elections, when an MP dies or resigns between general elections.

Parties have lists of people that they have decided would make good MPs, and your next step is to convince your party to put you on their list. They will ask lots of questions about your beliefs, and about your private life. Think hard, and tell them everything as honestly as possible. If they pick you, and then you ruin a campaign by having a guilty secret found out, you would never be forgiven.

Unless you can convince your local party to put you forward as their choice, you might be sent anywhere in the country to fight for a seat for your party, and you must be prepared to move there if you win it.

Seats far away from the House of Commons can mean that you spend up to a whole day every week travelling backwards and forwards on the train!

The party will allocate seats according to whether they are easy for it to win or not. Senior members of the party have to be given quite safe seats (areas where their party nearly always wins and is known to be the most popular), so that they are not suddenly knocked-out of Parliament. As a junior member on the list, you may well face a really difficult contest to begin with. If you win it, you'll have to work hard to make yourself really popular and well known so that your seat is safe in time for the next General Election.

CHAPTER THREE

ELECTIONS

General Elections come around at least every five years in the UK. To find the winning party in a General Election, all the votes in the country aren't just all added together as you might expect. Instead, voters are divided into geographical groups called constituencies. Local councils have the smallest constituencies, MPs the next smallest, and MEPs the largest. MPs each represent a constituency of about 90,000 people. This is quite cosy compared to other big countries – US senators represent nearly 600,000 people each, Germans 121,000, and French ones 102,000.

As people leave their home areas to settle in different places around the country, perhaps in big cities where it's easier to find work, boundaries have to be revised every now and again to keep the numbers in each area roughly equal. There's a special group called the Boundary Commission which does this every ten years or so in each part of the UK – England, Scotland, Wales and Northern Ireland.

Gerrymandering

In the past, corrupt governments have sometimes altered the boundaries of constituencies to make sure their side won. The first famous example of this was Governor E. Gerry of Massachusetts, USA, in 1812. His government drew up new boundaries in a strange shape to ensure that they would win the next election. The shape was a bit like a salamander (a sort of newt, but bigger). Angry Massachusetts voters nicknamed the new map 'Gerry's Salamander', or 'The Gerrymander' for short. Bending the rules like this is still known as 'gerrymandering' today. It's about the only thing poor old Governor Gerry is still remembered for!

Counting systems

Election results can also be changed by different ways of counting the vote. The UK currently uses a system called 'First Past the Post'. In this system every voter gets a single vote, and the candidate with the most votes wins the seat. If more than two parties are involved, this often means that most of the voters are not represented. For example:

Mr Pink – 38%
Mr Turquoise – 35%
Ms Purple – 27%

Mr Pink wins, even though 62% of the electorate voted *against* him! Whenever such results happen nationwide, Mr Pink's party can form a government, and the unpopular Mr Pink gets to be PM! Worse still, the UK system sometimes results in the party with the most votes *losing* the election! That happened to the Labour Party in 1951 and to the Conservative Party in early 1974.

For these reasons, MPs in opposition to the government are often unhappy with the electoral system. In the past, both Conservative and Labour MPs have promised to change to a fairer system. But once they are in power, it seems to be very hard for them to carry out their promises! The Liberal Democrats have always wanted a fairer system, but haven't won a general election yet. Perhaps you only mind about the system when you're on the losing side.

There are many arguments over what a replacement system would be like, but most of them agree on the principle of 'proportional representation'. In such a system representatives would be elected to the House of Commons in proportion to the number of votes their parties won.

Be a Prime Minister:
HOLD AN ELECTION!

The best way to test the different ways of running an election is with an experiment – try holding an election yourself!

WHAT TO DO

Ask your school if you can stage a 'mock' election – one that is only held for the fun of it. Advertise to see who wants to stand for election, and ask them to submit a one-page manifesto for the party (real or invented) that they represent. Research the different ways of counting votes and design a ballot paper (voting form) that uses all of them. On election day, the whole school (or perhaps just your year) comes to your ballot box (a special box which has a slit like a letterbox for inserting ballot papers) and fills out the ballot paper with their choices. Then count the votes according to the different systems and see if the results come out the same.

WHAT HAPPENS?

You'll probably find that different systems give different results. Which system do you think is the fairest?

Getting elected

To get elected as an MP, you'll have to make sure that the voters know who you are. Your party will be running a national campaign, making headlines, buying advertising space, and running election broadcasts on the telly. It's up to you and your constituency to make sure this message gets through in your area.

Party bigwigs will be very busy at election time, but try to get one of them to pay you a visit. This will give you a great chance to get your mugshot in the local paper. You can also try meeting up with local community groups to ask for their help on election day. Elections are often lost when people who agree with you don't bother voting. Make sure that your campaign touches on issues people feel really strongly about. Make sure they know how bad things could get if your rival wins. Otherwise, why will they bother switching off the telly and heading down to the polling station to vote?

Taking knocks

Sometimes you may find the battle between you and your rivals gets personal. If you were honest with your party about yourself, you should have nothing to fear. Try to ignore the sillier attacks on your character, and concentrate on talking about your party's manifesto – the important things it will try to do if you get into power. If you discover something terrible about one of your rivals, think carefully before telling the press – it always looks bad to be calling people names.

Whatever happens, don't ever lose sight of what you're standing for. If you stick to your own issues and work hard, the voters should respect you for it.

The Speaker

One MP is elected without having to go through all this trouble. The House of Commons has a chairperson called the Speaker, who is also an MP. A long time ago the major parties agreed that they would not compete for the Speaker's constituency, so she or he can be re-elected without contest. Speakers are trusted to be absolutely unbiased in the House of Commons, so it would be silly for them to battle with other candidates for their seat.

Election day

On election day you can cast a vote for yourself, and don't forget to round up all your friends and family to make sure they all get out and vote for you!

You might try to get onto local radio to remind people to vote for you. Voters are coming and going all day from the polling stations, but no one begins to count their votes until the polls close in the evening. Then an official called the Returning Officer sets their team to work counting the ballot papers.

Eventually the Returning Officer announces that they have been given a result. All the candidates gather behind the Returning Officer as they announce the results, in alphabetical order. If your constituency was a particularly interesting one, TV cameras may be there.

With a bit of luck, when the Officer reads your name, they will also read a very big number. Your team will cheer. If the number is bigger than anyone else's is, you've won! Your campaign team will be cheering like mad. You'll have to keep calm yourself though, because when you win you'll have to to make a short speech thanking everyone.

CHAPTER FOUR

WELCOME TO WESTMINSTER

You're in! Now what? Where do you go? Where do you sit? What do you do? Don't worry, your party will have lots of advice. Anyway, before you can even think about taking your seat in the House of Commons, you have to be sworn in. On the floor at the entrance to the House of Commons is a white line called the bar. You stand here and pledge your loyalty to the country. You still won't be a full member of the House until you've given your maiden speech, in which you introduce yourself to the other members. You are not supposed to say anything controversial in this speech, but as soon as you're finished, what you do in the House is up to you ... well, within limits.

The House has its own rules, just like school. If you approve of what an MP is saying in a speech, you can say 'Hear, hear!'. Long ago MPs used to shout 'Hear him, Hear him', but now they don't bother with the 'him'. You are not allowed to swear, or call other MPs liars, or mention the House of Lords. In practice, MPs often get around these rules. Winston Churchill used the words 'terminological inexactitude' to suggest another member was lying, and others have used the phrase 'economical with the truth'.

The rule against mentioning the House of Lords was meant to stop the two Houses blaming each other for problems. But MPs have found a way to talk about the House of Lords as well – they just call it 'another place'. You may amuse your fellow MPs with witty variations on the theme, but don't overdo it.

There are other rules every MP must obey, or else! When the House gets too excited, the Speaker will say 'Order, order!' If you don't quieten down, the Speaker may stand up. If they do this, every other MP has to sit down and shut up (even the PM), or the Speaker will 'name' them. Naming is a serious business – it's a bit like being suspended from school. Normally, MPs are never referred to by their real names while they are in the House of Commons. You have to use titles such as 'The honourable member for Piddletrenthide'. The Speaker only uses your name if you have done something wrong, to remind you that you are not living up to the standards of membership. If you ignore even that, the Serjeant at Arms and doorkeepers can throw you out into the street!

But perhaps the most important rule of all is to do with the Register of Members' Interests.

The Register of Members' Interests

The Register of Members' Interests isn't, as it sounds, a list of who likes gardening and golf in the House of Commons! It is a record of the money you make beyond your MP's salary. It is very important that you declare these earnings, including special treats people offer you, so that all the other MPs can check you're not being bribed. Some MPs are already rich, so they have to declare how their money is invested. The public may not be too bothered if you make a political mistake, or suddenly change your mind about an issue.

➤

But if you're caught campaigning for something that will secretly make you richer, your political career will be over, and you could even end up in prison!

What does an MP do?

Don't worry too much about all these rules though. As a young MP a lot of your time will be spent outside the House, looking after your constituency. Almost every MP sets aside a time in their weekly diary to meet concerned voters from their area. This is called a surgery. For some MPs, especially independents, representing their constituents' interests is the main reason for being an MP. Even career politicians like you, who secretly have their eye on higher things, should never forget their electorate. Remember that the next election is never more than five years away.

Debates

Every political party employs a few MPs to make sure that you're kept informed about the really important debates coming up in the Commons. These MPs are known as the Whips. Each week they distribute a booklet of next week's business. The Whips underline important items, underlining the really important ones three times. This means that your party expects you to vote on their side when at the debate. If you miss one of the really important ones, known as a 'Three Line Whip', you will be in very serious trouble with your party, regardless of your excuse.

You don't have to be in the House of Commons for the debate itself, unless you plan to speak or want to listen to the debate. However, you must be able to get to the House within six minutes of a vote being declared (called) so you can cast your vote on behalf of your party. Debates can be longer or shorter than expected, so don't try to guess what time voting will happen – your party may lose because you were too late to cast your vote! It only happens rarely, but sometimes the government has lost a vote because its MPs were caught napping. Don't delay for any reason – an MP once narrowly saved the day when he arrived dripping wet, wrapped in a bathtowel just in time to cast the deciding vote!

Voting in the House

Votes in the House of Commons are called divisions. When an item has been debated long enough, or when an MP persuades the House to agree to a vote, the Speaker asks those in favour to say 'aye', and those against to say 'noe'. The Speaker may suggest, just from the number of voices, that the 'ayes' or the 'noes' have it – meaning they are the winners. Most often, though, the House wants firm proof, and 'divides'. The exits from the division lobbies (rooms) are closed, after officials have checked that no one is inside already. The division bell rings throughout the Palace of Westminster, and policemen in the corridors shout 'Division'! MPs emerge from all sorts of nooks and crannies – offices, bars, meetings – and race to the division lobbies. There is great excitement – well, for those MPs who hear the call, there is.

One of the division lobbies is for the 'ayes' and the other is for the 'noes'. Make sure you know which is which! Exactly six minutes after the bell started ringing, the entrances to the lobbies are locked – if you're late, tough!

Then, the doors are unlocked again and the MPs walk out, one by one. They file out past a desk, where their names are recorded and a careful count is made of the numbers from each of the division lobbies.

On difficult votes, the atmosphere in the division lobby can be deadly serious. Some MPs are willing to sacrifice their popularity and career to vote against their own party's instructions because they feel strongly about the issue in question. If you ever think you're going to have to do this, you had better tell your constituency party first to make sure they agree with your stand – never forget, you are supposed to be representing them, after all.

Parliamentary quirks

London, and the River Thames especially, were so smelly before Victorian sewers were built that Parliament only sat (met) from February until July, and took the rest of the year off! Now Parliament disbands in summer and is recalled every autumn, with a Queen's Speech. The Queen takes to her throne in the House of Lords, and summons the MPs from the other House.

Recalling Parliament involves many odd rituals. The Queen sends a messenger called the Gentleman Usher of the Black Rod (Black Rod, to his friends). When Black Rod arrives at the House of Commons, the MPs slam the door in his face! He has to knock politely to beg their attention. This reminds everyone who is boss.

Then MPs listen to the Queen's speech to find out what the government plans to do over the next 12 months (the PM will have given her the list beforehand).

Next the Mace is placed in a special cradle on the table in front of the Speaker. The Mace is a symbol of the powers MPs have won from the Crown, and it is treated with great respect. The Mace must be in position before the House can begin work. An angry MP once interrupted a debate by lifting the Mace, but this isn't recommended – the Serjeant at Arms, who guards the Mace, has a very big sword!

Making laws

The government tries to achieve its goals by making laws. Sometimes these involve changes to existing laws, and sometimes they are completely new laws. The process for making a new law is quite complicated, since the law has to work well in our complicated society.

Take a few deep breaths, and we'll look at how it works... most of the time!

First, the government's attitude towards a problem, and its rough solution, is described in the Queen's Speech. Experts from the ministries write-up a first version of the new law called a draft Bill. The experts show this to the people who will be affected by the law, and make improvements where they can.

The draft law is now published as a White Paper, and a committee of MPs from all parties studies it carefully. Citizens and pressure groups can add their opinions here.

Once all the resulting comments have been taken into account, the law is drawn up as a Cabinet paper, and submitted to the Cabinet for approval.

If the Cabinet approves of all the changes made to the outline of the law given way back in the Queen's Speech, it passes the law on to the Treasury to see if the country can afford it.

If the Treasury says OK, the law goes to special lawyers called Parliamentary Counsels, who look at the law very carefully to spot any problems, however small, in the way the law is described.

At last, the law can finally be submitted to Parliament as a Bill. That's when the hard work really begins!

From Bill to Act

Each Bill has to have three 'readings' in the House. In practice, the hard work is done by committees of MPs between readings, but there can still be long debates in the Chamber itself. Sometimes the committee stages of a Bill take so long that it never arrives at its third reading. MPs opposing the government's plans know that this is where they stand most chance of sabotaging a Bill, or at least forcing the government to water it down a bit. The Bill might also be ambushed in the House of Lords. The Lords could once stop a Bill in its tracks by voting against it. Now the Lords can only refer a Bill back to MPs with suggested improvements. Still, this is a time-consuming process – to keep a Bill on schedule, the government often has to bargain and compromise with the Lords.

After the third reading, the Bill gets the thumbs-up from the Queen – called Royal Assent – and becomes a law. It is an Act of Parliament, or an Act. Everyone in Britain is supposed to keep an eye on new Acts, in case something they do has been made illegal. If they are arrested, even if they don't know about the new law, the courts will tell them it's their own fault! After all, there's plenty of warning – some Bills take months (or even years) to become Acts.

Don't despair, though! Not all Bills are controversial, and simpler Bills can get through much quicker than complicated ones. Edward Heath, a Prime Minister in the 1970s, managed to push one Bill the entire way through Parliament in just 17 hours – to rescue the struggling car company, Rolls Royce.

If you're finding it all a bit too much already, bad luck. You can't resign until the next General Election! A long time ago, people used to hate having to be Members, and were forced to come to the House. Resignation was banned, and the only way to leave was to apply to the PM to appoint you to a job that rules you out of membership. The Prime Minister can make you a Bailiff of the Chiltern Hundreds, which is a job without any duties or pay. Since these bailiffs cannot be MPs, you could then be released from the House! But you don't really want that do you? So, onwards and upwards then!

Be a Prime Minister:
PRACTICE YOUR SPEECH-MAKING SKILLS

One important skill you can practice before you get to the House of Commons is the art of making speeches. There's not much point talking to yourself in the mirror, though, so why not organize a debate at your school? Ask one of your teachers to help. Find a time when everyone who is interested can meet, and pick a topic that's in the news so that people are excited about it.

It will be easy to do some research – just read some newspapers and watch the news on TV. Make sure you find a nice, juicy controversial topic to get people interested enough to come along and hear you talk about it!

WHAT TO DO

Debates are held so that a vote can be held at the end to see who won. If you want to debate green issues, for example, you must pick a proposal, or 'motion' to debate, say, 'This House believes that Genetic Science will benefit humankind'. One or two speakers should prepare a speech defending the motion, and one or two attacking it. Advertise your debate throughout the school on noticeboards.

WHAT HAPPENS?

At the debate, the Chair (probably the teacher you recruited earlier) asks each speaker to deliver their speech. After that, the Chair will take questions and speeches from 'the House' (all the other people in the room). At the end, the main speakers can sum up their arguments, and everyone votes. Good luck!

What makes a successful speech?

In a debate, should you try to ignore some of your rivals' comments in the hope the audience will forget them, or is it better to deal with each and every point? Is it best to make a calm, logical argument, or to get passionate and emotional about your views? One PM, Disraeli, used to get so passionate when he made speeches that his false teeth popped out!

These pointers should help you to make a successful speech.

- Every speech needs to start with a powerful point to get the audience's attention, and end with another powerful point that they will remember. Some speechmakers like to make the bit in between as short as possible, others like to throw in a story or two, some facts and figures, and so on.

- The way you deliver your speech may be more important than the words themselves. People may get bored if you read a speech word for word from the page, so memorize your main points and just glance at your notes now and then to remind yourself where you're up to. (This is good practice for the House of Commons, too – reading speeches word-for-word is banned there.)
- Try to look at your audience instead of just mumbling into your notes. That way people feel you're really talking to them. It will also mean that you'll be able to see which parts of your speech go down well with your audience.
- Learn from experience. Remember, even when you lose a debate (and you probably will now and then), you should look back at it to see why it failed. What went wrong? Was there anything you could have done better?

CHAPTER FIVE

GETTING TO THE TOP

There are more than 650 MPs, but only one of you can be Prime Minister. How on earth are you going to compete with all the others? They are all very clever and good at getting their own way – just like you. But hang on. Some don't want to be PM, and others have left it too late to stand a chance. That probably narrows it down to about 400 rivals. Easy?

Well, however good your policies are and no matter how wonderful your intentions are, the real secret to becoming Prime Minister is making the great British public and the other members of your party love you. Basically, the name of the game is about making friends and influencing people!

How to win friends and influence people

See how many of these tactics you can manage, to get yourself noticed in Parliament.

- Always be a brave defender of your party, but don't just repeat what your leaders say. Find your own, serious reasons for believing in your party's manifesto and make sure everyone hears about your views. That way, people will be impressed by your thoughtfulness and you may gain a few supporters, as well as finding a better way for your party to explain its ideas to the people.
- Alternatively, rock the boat a bit. Get yourself a name for being a troublemaker who does whatever it takes to stand up for what you believe. Don't cause too much trouble though – you don't want to make too many enemies as you climb that ladder to success. You never know whose help you might need when you get there!

- Work hard. Volunteer to help out in the committee rooms. Force yourself to smile as you willingly work through piles of dull constituency paperwork. Oh, and don't forget to make sure everyone hears about how hard you are working by putting in some time at the party's offices where you can go on about your achievements to anyone who'll listen.
- Take care to understand how the House of Commons works. You can get caught out if you don't play by the rules, and even an apparently minor slip-up can make you look stupid and lose you potential supporters. For example, one MP did not realize that if he drew attention to the small size of his audience, the Speaker would have to call for a count. The Speaker interrupted his speech, called a count, and discovered that there were less than 40 MPs in the House. Since the House must have more than 40 MPs present to do anything, the hopeless MP's speech was cancelled and everyone was sent home early!

Finally, whatever you do, make sure you are brilliant at it. As soon as people realize what an asset you are to the party and how seriously committed you are, they'll start considering you for one of the more important jobs.

Wooing the media

However many brilliant political successes you manage to concoct, they won't count for much if you don't learn how to make the best of them. It's vital that you make sure your successes (and preferably not your failures) are reported in the media – on television, radio and in newspapers.

You can employ press officers who are experts at understanding what journalists need. These advisers are nicknamed 'spin doctors'. Newspapers and television programmes are always looking for good stories to tell. Your spin-doctors try to ensure that journalists can see your point of view, even leaking government secrets when it helps you. They also suggest ways of presenting the story which will help make the reader or viewer sympathetic to your opinion.

Getting in the news ....

Be crafty. Most newspapers only have one or two big headlines each day. Even if you have an important story for the media, it may be ignored if there's a bigger story around that day, like a major disaster or a big football match. Pick a time when there's nothing much else for them to talk about, and you could be famous for the day! Start small if you have to. It's a good idea to make sure that local papers in your constituency tell your voters all about your good deeds.

Be noticed!

The House of Commons has holidays, just like school. The long holiday in the summer is sometimes called the 'silly season'. This is because journalists get bored and write silly stories because there is so little political news about. If you have a story that can wait, release it in the silly season and you might make the front page, however boring it is really!

Manipulating the media

Some senior political journalists are allowed into the Lobby of the Houses of Parliament, where they may ask you embarrassing questions. They are called Lobby Correspondents. If you want to let them know about something, the Lobby can be a good place to find them. If you're trying to avoid difficult questions, there are ways of sneaking past them! You can even get into the House of Commons avoiding the Lobby, by entering from behind the Speaker's chair.

Sometimes you might want to let journalists know what's happening, but don't want them to reveal how they know about it. Ask if you can talk 'off the record'. This means they can discuss your story, but can't say who told them about it. This lets you test public reaction to an idea while avoiding the risk of making yourself look silly.

Working for the PM

So, thanks to all your efforts to get yourself into the news, more people are aware of your existence. Now you need to start wheedling your way into the corridors of power. If your party is in power (when it has most seats in the House), your main aim is to be given a job by your current Prime Minister. (Don't make it too obvious that what you're really after is their job, of course!)

The PM can give you various small jobs to do in the Commons itself, or you might be appointed to a ministry. It will be a while before you are senior enough to be in charge of a ministry, but you can become a junior minister to learn the ropes.

If you are in opposition (second most seats in the House), these jobs are a bit less exciting, but still important. You can't go to run a ministry, but you can join the Shadow Cabinet. This is the team of opposition MPs who might run the Government if their party won the next General Election.

What are ministries?

Ministries are departments of the state. The most senior people below the Prime Minister are the MPs who run each of the ministries. They are known as Secretaries of State. That is why, for example, the MP in charge of the Foreign and Commonwealth Office is known as the Foreign Secretary for short.

Some ministries are seen as more important than others are, but the PM has final say over which ministry comes top. The PM can even chuck out whole ministries, merge bits of them together, and create new ones.

The most important ministries

THE TREASURY – this is a bit like the government's accounts department.

THE HOME OFFICE – this is responsible for passports, the police and law and order.

THE FOREIGN OFFICE – this deals with other countries, helping trade, running embassies overseas, and booting out spies.

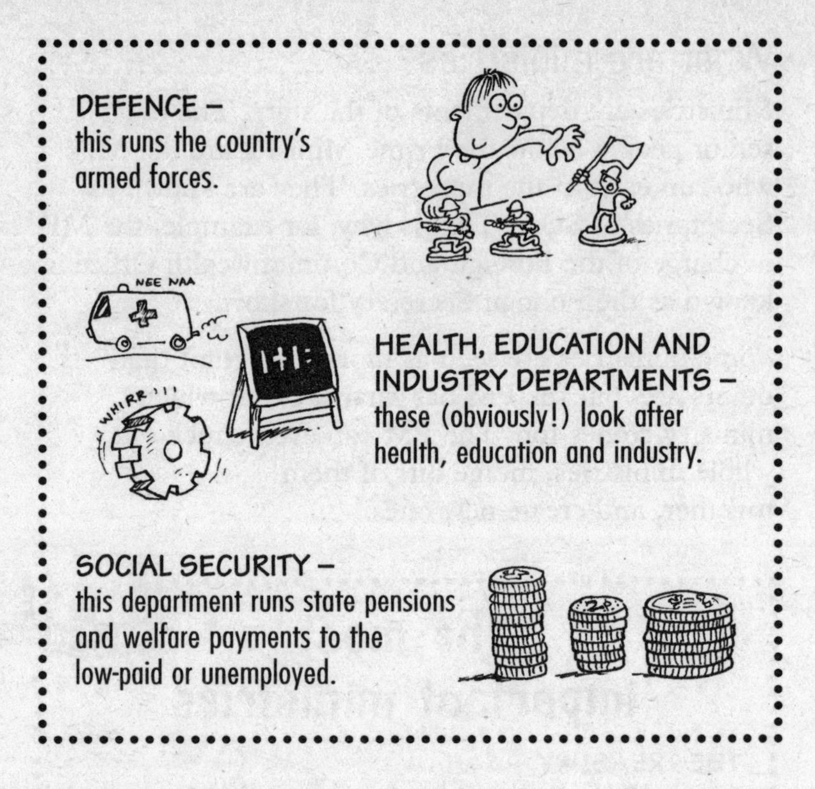

Perks and private secretaries ....

Ministers have very busy lives, but it's the best way to learn how to be PM, and you'll have a chance to impress your party with your talents. You'll spend a lot of your day with civil servants (the men and women who work in the ministry you're running). One of them will become your Private Secretary, and help you through the maze of the department – each one has a history almost as complicated as Parliament's.

You will also be given a government car and a driver to whisk you around London, and lots of secret documents in big, red briefcases (called 'boxes').

At the Foreign Office, top secret telegrams from your ambassadors are locked into little metal tubes, and only you possess the key!

Civil servants are paid by the state to do what their ministers tell them, but that's not the whole story. Civil servants have long careers, probably longer than yours will be – you're just another passing phase for them. Sometimes they can offer you expert advice. At other times, their advice might be intended to slow you down, especially if you are making a dramatic change. You'll have to learn to tell the difference between friendly advice, and advice intended to trip you up. If you don't, you'll either end up doing what your civil servants want, or you'll constantly be fighting them. Either of these could spell disaster for your career prospects.

Getting higher...

You're speeding ahead on the road to success. You've managed to get your ugly mug on the pages of various newspapers and done some radio and TV interviews as well. People recognize you on the street. (Well, some of them do.)

Your fellow MPs are impressed that you have taken responsibility and shown that you know how to use it. You have made friends with important people at home and abroad in your role as a Secretary of State, running a ministry. Influential people in the media and in your party are beginning to ask you when you're going to declare an interest in becoming Prime Minister.

At last you're ready to begin a serious campaign for the top job. Let's go!

It's your party

Each party has its own method of electing its leader. Of course, if you started your own party, you're probably the leader already. But if you're in one of the three main parties – Labour, Conservative, or LibDem – you'll have to follow their rules. Each of the big parties requires that you have the backing of a number of your party's MPs to become leader. The LibDems and Conservatives use a simple system in which each member has a vote and the candidate receiving the most votes wins. Conservative candidates have to win at least 50 per cent of the vote, so if there are three candidates and none reaches the target, the person with the lowest score drops out and the remaining two have another go. Labour has a slightly more complicated system, which divides votes into three sections. Members of the party count for a third of the vote, party MPs count for a third, and organizations connected to the party, such as trade unions, count for another third.

Let's say your party is in opposition. (Remember? This means they had the second highest number of votes at the General Election and sit opposite the government in Parliament.) You go to the party conference in the autumn determined to come away as party leader. All around you are fellow MPs and other party members who believe you can do it, like your style, and want to help you. All through the conference week you and your team work terribly hard to convince the necessary number of voters to support you. It will be the toughest fight you've had in your career – people you've grown up with as friends may decide to support your rival. Rumours from your past, true, semi-true and completely false, will sweep through the conference halls. But by now you're used to such difficulties, and you are equipped to deal with these problems.

Then it's over. The count is finished, a cheer goes up, your campaign team looks to you with big silly grins, and you realize that you have won! In the excitement of the next few seconds, remember to say some nice words to your rivals. You'll need their support as you campaign to become Prime Minister. You'll also need to take this opportunity to announce your big ideas (if you have any) to the nation. In opposition, you have the advantage that the public is bored of being lectured by the current Prime Minister, and you are now their most realistic alternative. With luck, and an enthusiastic party behind you, you're in a position to storm the heights of Westminster!

The General Election

Your first job will be to prepare a new manifesto for your party. Once the government calls a General Election (which happens every four or five years) you'll have just a few weeks to get your message across, so don't leave it to the last minute! Manifestos have to be carefully written, because there are always things happening in government that might make some plans impossible. You might find that you couldn't fulfil one of your promises without causing all sorts of new problems. For this reason it's best to be honest, and say only that you will do what you can. Never say never!

Staying in touch

When writing a manifesto, it's very important to know what the voters are thinking. There are people who specialize in measuring voters' opinions by taking big surveys, but often their results don't match the way people actually vote in an election. What are you going to do?
One answer is focus groups. These are small groups of people who volunteer to be quizzed on their reactions to a range of political issues.

Advertisers use focus groups to test whether a new product or advertising campaign will be popular. Many politicians think that focus group surveys are a good measure of public opinion. In the end, though, it's probably safest not to put too much importance on these surveys. If you do, you will be accused of 'government by focus group' – letting yourself be blown off-course by survey results, rather than sticking to your principles. Always remember that your job is not simply to do whatever the voters want, but also to show them why some unpopular things are necessary because they will be good for everyone in the long run.

Organize an army of MPs to be available to the media at all times – for interviews on TV, radio and in newspaper. Make sure the ones you choose are photogenic – silly hairdos and funky yellow flares are not going to inspire much confidence in the public. Projecting the right image is, as ever, all-important!

Use the media to get voters excited enough about your ideas to bother voting. Make sure that your party is getting a reputation for dramatic new plans – voters must be able to imagine how much better off they would be with you in charge!

If you call a press conference, watch out for hidden surprises, and answer only what you know about. If you have something special to say at a press conference, make sure you work it into every answer you give, regardless of the question. That way the voters will read your message – whichever piece of your statement the paper uses in its report.

Be a Prime Minister:
LEARN ABOUT 'SOUNDBITES'

There is not enough time or space for newspapers and TV news programmes to report politicians' entire speeches, so they just use snippets from them. Politicians have learned to use 'soundbites' – short sentences capturing the spirit of the speech which they use repeatedly. This is a clever way to get the media to report the message you want to get across. Soundbites are often a bit funny sounding, may lack verbs and can be serious or dramatic. Poetic, even. You'll get the hang of it if you follow this exercise.

WHAT TO DO

Look in your TV guide and work out which news programmes you are going to watch. Pick one on each channel. Grab a notepad and pencil, and listen hard. Sometimes the newsreader will tell you what an MP said, and sometimes you will see a clip of the MP saying it. Make a note of the words. Repeat this with different news programmes. You may find that some of the stories are similar, and use the same words from an MP's speech. Draw a line under the phrases you hear twice or more. You may have spotted a soundbite! If it's a really good one, you will have underlined it once for each news programme you watched. Start a collection, and you'll begin to see what makes a really good soundbite.

Be clever with the Civil Service, too. They'll be asking you for your plans in case you win. Knowing what to expect makes for less disruption, which civil servants detest! If you can soothe their worries about your radical plans, they'll tell their influential friends (newspaper editors, perhaps) that there's nothing to fear if you win. Handy, that!

It's the night you've been waiting for. People have been voting in the General Election all day. The election results come in as you gather with your campaign team. Everywhere you look, voters are swinging towards your party by the thousand. The number of seats your party has won is climbing, but then your rival party starts to catch up. Then more counts go your way – Phew! Suddenly, you've passed the halfway-mark! Your party controls more than half the seats in the House of Commons. Your head swells with pride, you've made it – you are, officially, the Boss!

CHAPTER SIX

YOU'RE THE BOSS!

First stop – Buck House. Yes, before you can start throwing your weight around you have to go to Buckingham Palace for the Queen to officially declare you Prime Minister. Then you go from the Palace to the second most famous home in the country – No.10 Downing Street! The staff in Downing Street will show you around, and you will meet lots of civil servants. Then you should call your chosen Cabinet members together and announce who will get the top jobs in your government. Now that you are PM, you can see all the top-secret papers that you weren't allowed to read as an opposition MP. There probably won't be any amazing surprises, but you'll want to know the whole story before you start making any changes.

Friendly politicians from other countries will be keen to congratulate you, so leave time for a chat with the American President and others. Your Cabinet should be raring to go, so make sure they have everything they need. You'll need to ensure that your team puts your election manifesto into practice, keep the peace when they disagree with each other, and help them solve problems. You will also have to attend lots of important international meetings with leaders of other powerful countries.

So, you have a country to run, a political party to lead, nosy journalists to keep happy, lots of MPs to look after, and, if that doesn't keep you busy enough, about half a million letters to answer each year!

Home, sweet home? ....

You'll be surprised at how much bigger your new home, 10 Downing Street, is than it looks from the front. Inside you'll find your private apartments, a meeting room for the Cabinet, and your own offices with around 50 staff!

Your press office is made up of people who are experts at dealing with newspaper and TV journalists. They keep an eye on the news for you, and act fast to stop any silly stories causing you too much trouble. Of course, as PM, you will have to expect trouble from the media sooner or later. About seven people handle your day-to-day business for you, answering letters, booking times into your diary, taking notes for you at meetings and that sort of thing. They even give a list of your appointments to your driver so that you don't have to tell him or her where to take you next.

You may have to decide whom to appoint to some very important jobs, such as the Archbishop of Canterbury. A special section at Downing Street studies candidates very carefully and helps to make these decisions easier for you. There is an office to handle the work you should be doing as an MP and to help you keep in touch with everything that happens in Parliament. You may have a couple of teams employed directly by your party to handle work that cannot be done by impartial civil servants.

It's up to you how many advisers you employ. Too few, and you'll rapidly fall out of touch. Too many, and your party will feel left out of decision-making. Not to mention that the media will tease you for not having any ideas of your own. These teams advise you how to steer your party to victory at the next General Election.

The Foreign Office

One of the most important parts of your job will be dealing with foreign leaders. The UK has trade links with a huge number of countries, and good relations are very important. Most of the time, your Foreign Secretary and the Foreign Office staff will be able to handle this business. But some meetings between nations will require your presence.

The Foreign Office should be one of your favourite departments, because you will spend a lot of time abroad meeting other leaders, and you will rely on the FO's information. Pick your Foreign Secretary carefully, to keep things friendly.

When Lord Home was Foreign Secretary in 1960, he had his ministerial 'red boxes' delivered to him while he was on a grouse shoot in Scotland. That wouldn't work too well these days!

Luckily for you, most of the hard work has normally been done before you even step onto your private jet. Officials in your ministries will have talked about doing a deal with similar officials in the other country. Your job is to chat to the foreign leader, perhaps make some small change to the deal, and to sign the paperwork. But every once in a while, you'll find that you have to do some quick thinking while you're there. You'll have to use all the skills you learned on your way up the career ladder!

Out there, somewhere!

The UK has to deal with America and four main groups of countries, and your approach to each of them will be different. The European Union, including the UK, France, Germany and Italy, and lots of smaller countries, is a sort of club. The UK has very important trade links with most EU countries, and it's vital to stay friends with them. Some EU members would like the EU to work more closely together. The UK has a reputation for resisting any attempts to unite the nations of Europe into a Union like the United States of America. Will you stick to this tradition, or do you think there is another alternative?

The Far East, including Japan, South Korea, Taiwan and China, has some of the most exciting economies in the world. Some are known as 'Tiger Economies'. Nobody knows how big China's market will become more involved in world trade. Business in the UK will be keen for you to help our country trade with these economies. You might be able to create jobs in Britain by encouraging these countries to invest their money here.

Many of the countries that used to be part of the British Empire also belong to a sort of club, called the Commonwealth. They all treat the UK as a close friend. Many use the English language, which makes it easy for UK businesses to trade with them. The largest Commonwealth country is India, with over a billion people. You will have to decide how much effort to put in to maintaining these links.

Poor countries that are not part of the Commonwealth form the last group. Britain helps these countries with a small Overseas Aid budget, especially when they have a crisis. Many of these countries owe money to big countries such as the UK, and some people think the UK should campaign to cancel this debt so the countries could spend the money solving their problems.

America is in a class of its own. It has the biggest economy in the world, the most powerful armed forces, and it is behind a lot of the world's most important organizations. Britain and America were allies in World War II, and Britain once seemed to be America's best friend in Europe. This was known as the 'special relationship'. It's up to you to find a balance between being America's friend, and getting closer to European powers such as Germany. Good luck!

What more advice can be offered to a person of your great stature? Not much, really. Never forget that you are the humble servant of the people who elected you. Try not to waste the rare opportunity this job gives you to do some good. After all, it may all be over after your four or five-year session is up!

CHAPTER SEVEN

BEYOND TOP DOG

All good things must come to an end. No one gets to be top dog for ever. Margaret Thatcher had the longest run as PM since World War II, and that was only 11 years (1979–1990). Most PMs are knocked off their perch by a General Election. Even if your party looks set to win all the elections for the next century, don't forget that your party is full of ambitious people like you who want to have a chance at the top job. Sooner or later, one of them will form a successful plot and replace you.

Of course, your head will be full of precious lessons when you leave Downing Street and return to the House of Commons. People will want to learn from you and use your skills in all sorts of ways. If you're clever, you might manage to see to it that the person who replaces you at No.10 is one of them. That way you never need miss the excitement because they'll be calling you up all the time to ask your opinion.

If you ever felt that you were underpaid for your brilliance, now is the time to make up for it. People will pay you several thousand pounds just to give after-dinner speeches at their clubs or business organizations. Publishers will want to print any book you care to write, and newspapers will fork out some cash (probably not much) for any comments you write for them.

You might find yourself invited back to Buckingham Palace one last time. Past PMs are usually made nobles by the Queen. This means you get to join the House of Lords. If you did a really great job there, you might just live to see a statue of yourself put up somewhere in Westminster.

Hope they don't make *your* ears too big!